Henry Draper

Cruise of School-Ship "Mercury"

1870 - 1871

Henry Draper

Cruise of School-Ship "Mercury"
1870 - 1871

ISBN/EAN: 9783337319991

Printed in Europe, USA, Canada, Australia, Japan

Cover: Foto ©berggeist007 / pixelio.de

More available books at **www.hansebooks.com**

Department of Public Charities and Correction.

CRUISE

OF

SCHOOL-SHIP "MERCURY"

IN

TROPICAL ATLANTIC OCEAN.

1870–1871.

NEW YORK:

THE NEW YORK PRINTING COMPANY,

Nos. 81, 83, and 85 Centre Street.

1871.

REPORT.

DEPARTMENT OF PUBLIC CHARITIES AND CORRECTION,
CORNER OF THIRD AVENUE AND ELEVENTH STREET,
NEW YORK, September 12, 1871.

Hon. A. OAKEY HALL, *Mayor:*

SIR—The Commissioners of Public Charities and Correction respectfully submit a report of the practice-cruise of the school-ship Mercury, from her anchorage ground at Hart's Island to Sierra Leone, and from thence *via* St. Thomas to this port, during the winter of 1870–71.

But, before narrating the history of the cruise, it is proper to state more specifically than in the annual report the reasons which induced the Commissioners to establish a nautical school, and to describe the character of the boys who are assigned to it. With the increase of the population of the city there had been a greatly increased number of boys committed to the care of the Commissioners by the magistrates, for slight misdemeanors and vagrancy. Others, and in large numbers, had been committed by their parents as incorrigible, or because of evil associates, who were leading them to ruin. These boys were at first sent to the Industrial Schools on Hart's Island, but the number increased so rapidly that the Commissioners were embarrassed as to the disposition which should be made of them.

They could not without a long probationship be recommended as apprentices, because of their wayward and reckless character, nor could they be discharged without the probability that they would again become vagrants, or fall into their former wicked associations. Under these circumstances it was deemed expedient to establish a nautical school, as well for the purpose of relieving the department of a constantly increasing number of unruly boys, as of providing for them a sure and honest means of livelihood suited to their adventurous spirit.

When the school was established it was contemplated from considerations of economy to teach them the rudiments of seamanship while the ship was at anchor in the harbor, and by occasional trips of short duration at sea, during the summer months; and this method was further commended by the consideration that the boys would be under the direct and constant supervision of the Commissioners.

But the experience of a few months demonstrated that the only effectual mode of instruction is the continuous handling of a ship at sea, and that the manifold duties of a thorough seaman can only be learned by actual service.

The cruise, of which the following papers constitute the report, was made in pursuance of these convictions and for the purpose of qualifying the boys for immediate entrance on their return into the service of the navy or mercantile marine. In this respect the result of the cruise was highly satisfactory, for of the crew of two hundred and fifty-eight boys over one hundred were, in the opinion of the captain, capable on the return of the ship of discharging the duties of ordinary seamen.

It is proposed to apply to Congress at its next session for authority to be vested in the Secretary of the Navy to discriminate in enlistments in favor of boys who have been educated in school-ships.

The exploration of the ocean has become an object of deep interest to governments as well as to men of science. The United States Coast Survey, under the wise direction of Professor Bache, and now of Professor Pierce, has for many years been diligently engaged in adding to the limited information possessed of deep-sea soundings, temperatures, and currents, and more recently the British Government, under the supervision of Dr. Carpenter, has solved several problems of scientific importance, added largely to the stock of general knowledge, and made discoveries which will be of practical benefit to commerce and navigation.

With the hope that the cruise of the Mercury might be made to contribute something of value to science, Captain Girand was directed to obtain a series of soundings on the line of or near the Equator, from the coast of Africa to the mouth of the Amazon, to observe the set of the surface currents, and the temperature of the water at various depths. The following is a copy of his letter of instructions:

DEPARTMENT OF PUBLIC CHARITIES AND CORRECTION,
NEW YORK, December 13, 1870.

SIR—You will proceed to sea, in command of the Mercury, on the 16th instant, by way of Montauk Point, and cruise between the latitudes 10° north and 5° south of the equator, and of longitude from 5° east to 45° west of Greenwich, until the 10th of March, when you will return to your anchorage at Hart's Island, unless on your approach to the coast you shall deem it prudent, from stress of weather, to enter at Sandy Hook.

You will keep at sea as much as possible, going into port only to obtain water and provisions or for repairs. As often as may be practicable, you will forward to this office detailed reports of the health of the officers and crew, and the progress made by the boys in seamanship and in their studies at school.

The routine duties of the ship, as laid down in the rules for the education of the boys in seamanship, will be strictly adhered to.

You will, as often as may be practicable, take soundings of the bed of the Atlantic, on or near the line of the Equator, from the coast of Africa to the mouth of the Amazon.

You will also frequently observe and record the temperature of the ocean at the surface and at twenty, fifty, one hundred, and two hundred fathoms, and obtain specimens of water at those depths, which you will cause to be evaporated on board, and the solid matter analyzed, or you will bring the specimens home in bottles provided for the purpose.

You will also obtain sea-plants from as great depths as may be practicable, together with animalculæ and other minute forms of animal life, which you will preserve or cause accurate drawings thereof to be made, for which purpose you will be furnished with a microscope of large powers.

An accurate knowledge of the set and velocity of currents is of great importance to commerce. You will be pleased to note and carefully determine the direction of all currents with which you may fall in.

A favorable opportunity will be afforded to note with approximate accuracy their rate and direction when you are engaged in taking your deep-sea soundings from an open boat.

It is desirable that the misplaced reliance of many navigators on dead-reckoning may be demonstrated, and to this end you will frequently, when you have determined your position by astronomical observation, throw your registering log and let it remain in the water for twenty-four or forty-eight hours, or until you have taken another observation, when you will compare the progress you have made by dead reckoning with your actual progress, as determined by observation.

While the chief object of the cruise of the Mercury will be to perfect the boys under your command in seamanship, the Commissioners indulge the hope that, by the careful observation of yourself and officers, the interests of commerce may be promoted, and the cause of science advanced.

Very respectfully,

ISAAC BELL,
President.

To Captain PIERRE GIRAUD,
Commanding School-ship Mercury.

The ship sailed from Hart's Island on the 20th December, 1870, and, after stopping at Madeira and the Canary Islands, arrived at Sierra Leone on the 14th of February.

On the 21st of February she left Sierra Leone, and on the 22d Captain Giraud commenced his deep-sea soundings, which he continued in nearly a straight line to the Island of Barbadoes, a distance of about

2,800 miles. His soundings range from 500 fathoms to 3,100 fathoms, or 3½ miles, from which depths he brought up in most instances specimens of bottom. The current observations will be of value in the navigation of the South Atlantic, and the ascertained temperatures of the water at from 200 to 500 fathoms are confirmatory of the theory that a cold current from the Poles underlies the surface-waters of the tropical seas.

The daily meteorological observations, the direction and velocity of currents, and the temperatures of the ocean at various depths, are set forth in his report, which is herewith annexed. These papers, together with the specimens obtained from the bed of the ocean and of seawater, were placed in the hands of Professor Henry Draper of the New York University for examination.

Professor Draper's report, which is herewith submitted, contains precise tabulated statements of the meteorological observations of the voyage, of the direction and velocities of the currents, and of the temperatures at the several depths obtained, together with an analysis of the specimens of water from various depths. The report also contains a diagram of the bed of the Atlantic, from Sierra Leone to Barbadoes, as established by the soundings, and is replete with interesting disquisitions on all the questions of deep-sea explorations. The specimens of animal life which were obtained from the bottom were forwarded for examination by Professor Draper to Dr. Carpenter. He reports that they are the ordinary forms of deep-sea foraminifera. With the specimens Professor Draper also sent a table of the temperatures, and in respect to them Dr. Carpenter remarks, "that they are of great interest, and especially those at one hundred and two hundred fathoms. They show," he continues, "how thin is the surface stratum affected by the gulf-stream, or by direct solar radiation. The sudden drop," he remarks, "in the temperature at two hundred fathoms, between 17° 46' west longitude, and 19° 36' west longitude, and the continuation of this reduction with the increased westing as far as 50° 38' is a very curious phenomenon, and I cannot help connecting it with some great oceanic movement, especially as at 68° 47' west longitude, and at 83° the higher temperatures reappear. I trust that hereafter much attention will be given to this point."

The conclusion at which Professor Draper has arrived, from a careful examination of the results obtained, is that there exists, all over the bottom of the Atlantic and Caribbean Sea, a stratum of cold water, and that the cruise of the Mercury must be considered as offering confirmatory proof of the correctness of Dr. Carpenter's theory, drawn from the cruises of the English exploring vessels, that there is a general movement of the lower waters of the Atlantic towards the Equator, and a corresponding flow of the surface-waters towards the Poles.

The object of the cruise was, as has been stated, to perfect the boys in seamanship. The scientific explorations were incidental and subsidiary, but enough has been accomplished to encourage the hope that the practice voyages of the Mercury may be made to contribute to the stock of knowledge, and help to explain the mysteries of the Great Deep.

The experiment of the school-ship, as a Reformatory, has thus far been satisfactory. There is reason to believe that it is the most effective mode to reclaim erring boys, whose errors, caused by the love of adventure, by evil associations, or ungovernable tempers, are fast impelling them to ruin. Brought under the inflexible discipline of a ship in actual service, they are taught in a few months the duties of a profession, which directs and gratifies their love of adventure, and provides for them the means of an honest and useful livelihood.

Respectfully submitted.

ISAAC BELL,
OWEN W. BRENNAN,
JAS. B. NICHOLSON,
ALEX. FREAR,
JAMES BOWEN,

Commissioners of Public Charities and Correction.

REPORT

TO THE

Commissioners of Public Charities and Correction

OF THE

CITY OF NEW YORK,

ON THE CHEMICAL AND PHYSICAL FACTS COLLECTED FROM THE

DEEP SEA RESEARCHES

MADE DURING THE VOYAGE OF THE

NAUTICAL SCHOOL-SHIP "MERCURY,"

UNDERTAKEN BY THEIR ORDER IN THE

TROPICAL ATLANTIC AND CARIBBEAN SEA,

1870-71.

By HENRY DRAPER, M.D.,

Professor of Analytical Chemistry and Physiology in the University of New York.

REPORT.

The Commissioners of Public Charities and Correction, having submitted to me for examination certain documents and specimens connected with the voyage of the nautical school-ship Mercury, I have the honor to make thereupon the following report:

Much attention has recently been given to deep-sea researches, in consequence of the investigations made by the United States Government on its coast, and by Dr. Carpenter, Mr. Gwyn Jeffreys, and Professor Wyville Thomson, in the North Atlantic and the Mediterranean Sea. Not only have many of the facts so ascertained been corroborated by this voyage of the Mercury, but the Commissioners, by authorizing it, have also added much that is new and interesting to our knowledge of the physical condition of the deep sea.

The voyage of the Mercury may be divided into three stages: 1st, from New York, by way of the Madeira and Canary Islands, to Sierra Leone; 2d, from Sierra Leone, through the tropical Atlantic, to Barbadoes; 3d, from Barbadoes, through the Caribbean Sea, to the north of Cuba, and thence, along the coast of the United States, back to New York.

The chief scientific interest of this voyage is connected with its second and part of its third stage.

The ship left New York on December 20, 1870. She reached Funchal (Madeira Islands) on January 17, 1871 ; Las Palmas, in the Canaries, on January 24, and Sierra Leone on February 14. On the 21st of that month she sailed for Barbadoes, and reached that island on the morning of March 17. She sailed from Barbadoes on March 24, and arrived in New York on April 21.

It is necessary to give these particulars, with a view of indicating that the voyage was made in the winter season of the year, the season of the lowest air temperatures. The mean temperature of the air at 8 P.M., while at Funchal, was 70⅖°, the maximum being 76° and the minimum 68°. While at Las Palmas it was 63°, the maximum being 66° and the minimum 57°. While at Sierra Leone it was 83°, the maximum being 85°, the minimum 79°. In the passage across the Atlantic, from Sierra Leone to Barbadoes, the mean temperature was 78¾° ; the maxima were on leaving the African and on approaching the American coast. For nearly a fortnight, during the mid-passage, the variations were included between 76° and 79°.

In this report the degrees of temperature are according to Fahrenheit's scale. Though the thermometer was observed regularly, three times each day, at midnight, at noon, and at 8 P.M., I have adopted the latter only, since it is recognized by meteorologists that the temperature at 8 P.M. very closely approaches the mean for the entire day.

As regards the barometer, the following table will show its height, as determined by three observations each day during the passage from Sierra Leone to Barbadoes :

Table I.

Barometric Observations from Sierra Leone to Barbadoes.

DATE.	MIDNIGHT.	NOON.	8 P.M.
Feb. 21	29.64	29.66	29.60
Feb. 22	29.70	29.68	29.70
Feb. 23	29.76	29.70	29.74
Feb. 24	29.70	29.68	29.66
Feb. 25	29.68	29.68	29.70
Feb. 26	29.74	29.74	29.74
Feb. 27	29.76	29.76	29.76
Feb. 28	29.74	29.68	29.64
March 1	29.68	29.66	29.66
March 2	29.72	29.68	29.68
March 3	29.72	29.70	29.70
March 4	29.72	29.74	29.72
March 5	29.72	29.73	29.74
March 6	29.80	29.78	29.78
March 7	29.78	29.78	29.78
March 8	29.78	29.76	29.76
March 9	29.79	29.79	29.78
March 10	29.76	29.76	29.76
March 11	29.77	29.73	29.73
March 12	29.74	29.76	29.76
March 13	29.78	29.76	29.76
March 14	29.82	29.81	29.82
March 15	29.80	29.80	29.74
March 16	29.78	29.78	29.80
Mean of all the observations	29.745	29.733	29.730

From this it will be seen how small the barometric variations were. Examining, for instance, those of 8 P.M., the minimum was only $\frac{13}{100}$-inch below, and the maximum $\frac{8}{100}$-inch above the mean. It may also be remarked that, in a general manner, the pressure of the air increased on nearing the American coast.

On leaving the African coast the atmosphere was hazy, as is usual in that region. On March 5 the north-east trade winds were struck, and they accompanied the ship to Barbadoes. Previously to reaching them her mean daily progress had been only 57 miles; after that time it was 187 miles.

The ocean currents, as reported on the days when other important observations were made, are shown by the following table:

TABLE II.

Direction and Velocity of Currents between Sierra Leone and Barbadoes.

DATE.	DIRECTION.	VELOCITY.	DATE.	DIRECTION.	VELOCITY.
Feb. 23.......	S.W.	½ knot.	March 4.......	S.	½ knot.
" 25.......	S. S.W.	½ knot.	" 10.......	S.W.	½ knot.
" 26.......	S. by W.	½ knot.	" 11.......	S.W. by W.	½ knot.
" 27.......	S	½ knot.	" 13.......	W. S.W.	½ knot.
" 28.......	S. by W.	½ knot.	" 14.......	W. S.W.	¼ knot.
March 3.......	S. S.W.	¾ knot.	" 15.......	W.	½ knot.

TABLE III.

Direction and Velocity of Currents in the West India Seas.

DATE.	DIRECTION.	VELOCITY.	DATE.	DIRECTION.	VELOCITY.
April 4.........	W.	½ knot.	April 8.........	W. N.W.	½ knot.
" 5.........	W.	¾ knot.	" 9.........	W.	¼ knot.
" 6.........	W. by N.	½ knot.	" 11.........	S. W.	¾ knot.
" 7.........	W. by N.	½ knot.	" 13.........	N. N. W.	½ knot.

During the second stage of her voyage the ship's track was over a parallel included substantially between the eleventh and thirteenth degrees of north latitude. She obtained soundings at various depths, from 290 to 3,100 fathoms, and on eleven occasions specimens from the bottom. In the third stage of her voyage, while in the West Indian seas, three other bottom-specimens were secured. The apparatus used was Brooks' detaching apparatus, with two thirty-two pound shot. The sounding-line was cotton cord, one-seventh of an inch in diameter. All the soundings were made from a boat. On March 6, the maximum depth of 3,100 fathoms in the Atlantic was reached, but on reeling in, the line unfortunately parted, and 2,200 fathoms were lost. This is one of the deepest accurate soundings ever made.

Samples of water from the surface, and also from depths varying from two hundred to five hundred fathoms, as shown in Table IV., were procured. The temperature of the sea, both at the surface and at those depths, was ascertained. The samples of water were collected in the usual apparatus, a metal cylinder, presently to be more particularly described, having at its bottom and top valves opening upwards. By this contrivance as water obtained from a great depth is drawn toward the surface, it and its included gases have liberty to expand, the excess escaping through the upper valve. If such provision were not made, the cylinder, no matter how strong it might be, would unavoidably be burst open. It should be borne in mind that the compressibility of water is about $\frac{1}{22000}$ for each atmosphere of pressure it sustains, and at depths such as were here reached the pressure was about 1,250 pounds per square inch, or more than eighty atmospheres.

There are certain precautions which must be attended to in the use of this collecting cylinder. These more particularly refer to securing the perfect action of its valves. It is intended that these valves should remain open during the entire period of the descent of the cylinder in the sea, and remain closed during its ascent, except in so far as the upper one may be opened by the interior pressure to allow the excess of included water and its dissolved gases to escape. Obviously, however, these conditions may be interfered with by a variety of accidental causes, such as the adhesion of the valves by verdigris or other impurity, or by the cylinder assuming an inclined instead of a vertical position.

The constitution of the water as it exists at great depths is not correctly represented by the samples thus obtained. A considerable portion of the gases dissolved therein may escape, as just stated, under the relief of pressure as the cylinder is drawn toward the surface, and hence examinations of such samples, as regards their gaseous ingredients, are liable to be deceptive. The low temperature and great pressure of these deep strata, moreover, increase the solvent power of the water over gases, and this power is diminished as the cylinder is brought into the warmer strata above, and into the open air. Even the saline ingredients will suffer disturbance when they are held in solution by gases that will thus escape. For instance, this is the case with carbonate of lime. No method has hitherto been practised which furnishes a means of obtaining samples of sea-water from great depths with their true constitution undisturbed, though obviously an apparatus might be devised which would accomplish that purpose.

As thus procured the specimens of water were preserved in well-corked glass bottles with sealing-wax on the cork, until submitted to me for examination. These samples are fifty in number, divisible into two groups, surface and deep ones, from each locality. In quantity they vary from four to sixteen ounces.

I have determined the specific gravities of these specimens and most of them are inserted in Table IV. To insure correct results several precautions must be taken. The difference of density between water collected at the surface and that from great depths is so small that a slight variation in temperature is sufficient to mask it completely. Hence it is only in laboratories, where means can be used to provide against temperature variations and where balances of precision can be employed, that accurate results can be obtained. It will, therefore, be understood that in the experiments upon which the following table has been constructed such precautions have been carefully attended to. It may be remarked that on board ship, where, on account of the motion, balances cannot be applied, the hydrometer must be resorted to, but any conclusions drawn from its indications should be accepted with much reserve. It is difficult to read its scale with exactness, and it is almost impossible under such circumstances to secure the proper temperature conditions. In some of the more interesting instances the variation of a single degree in the temperature would lead to a conclusion in direct opposition to the true one.

TABLE IV.

Specific Gravity of Samples of Sea-water at 75°.

DATE.	LATITUDE.	LONGITUDE.	DEPTH.	SPECIFIC GRAVITY.
Feb. 25	9° 14'	17° 09'	Surface	1026.72
Feb. 25	9° 14'	17° 09'	200 fathoms	1026.68
Feb. 26	10° 04'	17° 33'	Surface	1026.42
Feb. 26	10° 04'	17° 33'	200 fathoms	1026.50
Feb. 27	10° 42'	17° 46'	Surface	1026.72
Feb. 27	10° 42'	17° 46'	200 fathoms	1026.66
March 1	11° 35'	18° 20'	Surface	1026.53
March 1	11° 35'	18° 20'	200 fathoms	1026.53
March 2	11° 39'	18° 33'	Surface	1026.80
March 2	11° 39'	18° 33'	200 fathoms	1026.80
March 3	11° 35'	19° 35'	Surface	1026.76
March 3	11° 35'	19° 35'	200 fathoms	1027.00
March 4	11° 06'	21° 55'	Surface	1026.92
March 4	11° 06'	21° 55'	200 fathoms	1026.77
March 10	12° 38'	42° 31'	Surface	1027.28
March 10	12° 38'	42° 31'	200 fathoms	1027.32
March 11	13° 02'	44° 51'	Surface	1027.10
March 11	13° 02'	44° 51'	200 fathoms	1027.28
March 12	13° 06'	50° 38'	Surface	1026.58
March 12	13° 06'	50° 38'	200 fathoms	1026.81
March 13	13° 08'	53° 48'	Surface	1026.72
March 13	13° 08'	53° 48'	420 fathoms	1026.88
March 15	12° 55'	56° 46'	Surface	1026.52
March 15	12° 55'	56° 46'	100 fathoms	1027.03
April 4	17° 13'	67° 29'	Surface	1026.83
April 4	17° 13'	67° 29'	400 fathoms	1026.87
April 5	17° 08'	68° 48'	Surface	1026.78
April 5	17° 08'	68° 48'	200 fathoms	1026.83
April 6	17° 09'	71° 47'	Surface	1026.87
April 6	17° 09'	71° 47'	(?)	1026.79
April 7	17° 27'	74° 33'	Surface	1026.88
April 7	17° 27'	74° 33'	300 fathoms	1027.03
April 7	17° 27'	74° 33'	400 fathoms	1027.16
April 8	18° 11'	76° 00'	Surface	1026.83
April 8	18° 11'	76° 00'	200 fathoms	1026.83
April 9	19° 12'	77° 54'	Surface	1027.04
April 9	19° 12'	77° 54'	200 fathoms	1026.96
April 10	20° 17'	80° 35'	Surface	1027.56
April 10	20° 17'	80° 35'	100 fathoms	1027.48
April 11	20° 48'	83° 02'	Surface	1027.08
April 11	20° 48'	83° 02'	200 fathoms	1027.04
April 13	23° 19'	84° 17'	Surface	1027.16
April 13	23° 19'	84° 17'	100 fathoms	1027.20
April 13	23° 19'	84° 17'	200 fathoms	1027.16

The temperatures under the sea-surface were obtained by a self-registering thermometer of the form known as Six's construction. It had no provision to protect it from variations of pressure. Compensa-

2

18

tion for these variations must be determined in the case of each individual instrument, for the amount of this error will vary with the varying thickness of the glass, its form, and its power of resisting compression. In the experiments made by Dr. Miller on self-registering thermometers for deep sea sounding, published in the report of the Meteorological Committee of the Royal Society for 1869, it is shown that certain unprotected thermometers submitted to a pressure of two and one half tons per square inch in a hydraulic press, though made with bulbs of unusual thickness, would indicate temperatures from $6\frac{4}{10}°$ to $8\frac{3}{10}°$ too high, and in other experiments when the pressure was raised to three tons on the inch the error was $11\frac{1}{2}°$. In these instances, however, a part of the rise, perhaps as much as $1\frac{1}{2}°$, was due to the heat disengaged from the water itself in the act of compression. In most of the subjoined observations, the depth not exceeding 200 fathoms, the pressure upon the thermometer was one quarter of a ton per square inch.

In the apparatus used on board the Mercury, the thermometer was inclosed in the water-collecting cylinder. It was, as in Six's form, alcoholic, the reservoir consisting of a tube about five inches long and one-third of an inch in diameter, made of pretty substantial glass, and, though the influence of pressure upon it has not yet been determined experimentally, there is reason to suppose that at the depths in question the error would not exceed one or two degrees. No index error was found on comparing this instrument with a standard Kew. The water-collecting cylinder consisted of a copper tube thirteen inches long and one and three quarters in diameter, weighted with a hollow cone of lead at its lower extremity. The valves above and below were one inch in diameter. In what may be called the front of the cylinder there were inserted strips of plate glass, through which the indications of the thermometer might be read without removing

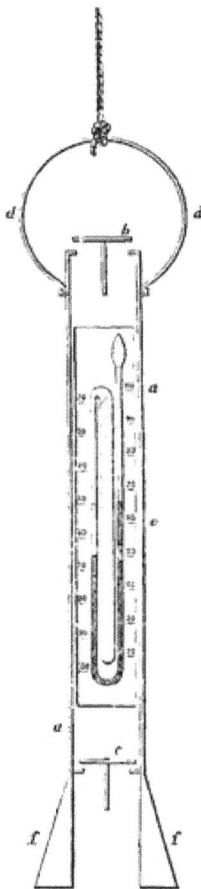

SECTION OF WATER-COLLECTING CYLINDER AND THERMOMETER.

it from the cylinder. The glass was protected from injury by brass rods.

In the above figure *aa* is the copper cylinder, *bc* the two valves, *dd* the handle for connection to the sounding line, *e* the Six's thermometer, *ff* the weight at the bottom.

TABLE V.

Air Temperatures between Sierra Leone and the Florida Capes.

DATE.	TEMPERATURE OF AIR AT			DATE.	TEMPERATURE OF AIR AT		
	Midnight.	Noon.	8 p.m.		Midnight.	Noon.	8 p.m.
Feb. 21....	79°	84°	81°	Mar. 14....	78	81°	81°
" 22....	79°	84°	83°	" 15....	74°	79°	77°
" 23....	80°	83°	79°	" 16....	79°	85°	85°
" 24....	78°	81°	79°	" 24....	86°	87°	85°
" 25....	78°	84°	80°	" 25....	79°	82°	82°
" 26....	77°	82°	76°	" 26....	80°	86°	84°
" 27....	74°	82°	77°	" 27....	81°	86°	83°
" 28....	74°	78°	79°	April 3....	82°	84°	84°
Mar. 1....	75°	80°	78°	" 4....	80°	83°	82°
" 2....	73°	80°	78°	" 5....	80°	81°	83°
" 3....	74°	80°	76°	" 6....	81°	81°	87°
" 4....	75°	77°	77°	" 7....	84°	84°	81°
" 5....	74°	80°	76°	" 8....	82°	82°	86°
" 6....	74°	79°	76°	" 9....	83°	84°	88°
" 7....	74°	79°	79°	" 10....	84°	84°	80°
" 8....	73°	78°	77°	" 11....	82°	84°	82°
" 9....	74°	78°	76°	" 12....	82°	85°	84°
" 10....	75°	80°	77°	" 13....	80°	84°	84°
" 11....	76°	81°	81°	" 14....	80°	86°	78°
" 12....	76°	81°	80°	" 15....	80°	86°	78°
" 13....	78°	84°	81°				

It will be remarked that in the foregoing table the temperatures are given for midnight, noon, and 8 P.M. But as the soundings were usually taken at other hours, more commonly at 3 P.M., I give in the following table the temperatures observed at those hours, both at the surface and at the depths specified.

TABLE VI.

Temperature of the Air, of the Sea-surface, and of the water at various depths.

DATE.		Hour.	Latitude	Longi-tude.	TEMPERATURE OF			
					Air.	Water Sur-face.	Water at various Depths.	
Feb.	23...............	2 P.M.	8° 50′	15° 47′	79°	78°	200 fath.	54°
"	25...............	3 P.M.	9° 15′	17° 12′	78°	79°	200 fath.	58°
"	26...............	3 P.M.	10° 05′	17° 35′	76°	74°	200 fath.	60°
"	27...............	3 P.M.	10° 42′	17° 46′	78°	77°	200 fath.	60°
"	28...............	3 P.M.	11° 24′	18° 09′	77°	76°	200 fath.	53°
March	1...............	....	11° 26′	18° 20′	80°	77°	200 fath.	53°
"	2...............	....	11° 39′	18° 33′	82°	77°	200 fath.	53°
"	3...............	3 P.M.	11° 35′	19° 33′	77°	76°	200 fath.	53°
"	4...............	3 P.M.	11° 06′	21° 55′	76°	77°	200 fath.	52°
"	6...............	3 P.M.	11° 32′	29° 13′	76°	75°	....	....
"	10...............	2 P.M.	12° 38′	42° 31′	83°	76°	200 fath.	50°
"	11...............	2 P.M.	13° 03′	44° 51′	84°	76°	200 fath.	51°
"	13...............	11 A.M.	13° 06′	50° 38′	80°	75°	200 fath.	51°
"	14...............	2 P.M.	13° 08′	53° 48′	80°	79°	400 fath.	47°
"	15...............	2 P.M.	12° 55′	56° 46′	81°	80°	100 fath.	62°
April	4...............	....	17° 13′	67° 29′	80°	84°	100 fath.	70°
"	4...............	3 P.M.	17° 13′	67° 29′	80°	84°	400 fath.	48°
"	5...............	3 P.M.	17° 08′	68° 48′	84°	84°	200 fath.	59°
"	6...............	....	17° 09′	71° 47′	82°	84°	200 fath.	62°
"	6...............	4 P.M.	17° 09′	71° 47′	82°	84°	500 fath.	48°
"	7...............	....	17° 27′	74° 33′	82°	84°	300 fath.	54°
"	7...............	4 P.M.	17° 27′	74° 33′	82°	84°	400 fath.	50°
"	8...............	3 P.M.	18° 11′	76°	85°	85°	200 fath.	62°
"	9...............	4½ P.M.	19° 12	77° 54′	84°	85°	200 fath.	62°
"	11...............	4 P.M.	20° 48′	83° 01′	84°	85°	200 fath.	63°
"	13...............	....	23° 19′	84° 17′	84°	86°	100 fath.	72°
"	13...............	5 P.M.	23° 19′	84° 17′	84°	86°	200 fath.	62°

For a proper appreciation of the conclusions to be drawn from these tables, it is necessary to point out the facts indicated by the soundings of the ship. For the sake of brevity I restrict these remarks to the second stage of the voyage.

Parting from the African coast, the bed of the ocean sinks very rapidly. A couple of degrees west of the longitude of Cape Verde the soundings are 2,900 fathoms. From that point the mean depth across the ocean may be estimated at about 2,400 fathoms, but from this there are two striking departures—first a depression, the depth of which is 3,100 fathoms, and second an elevation, at which the soundings are only 1,900—the general result of this being a wide and deep trough on the African side, and a narrower and shallower trough on the American. It may be that this peculiarity is a result of the river distribution on the two continents respectively, there being, with the exception of the Senegal and Gambia, no important streams on the African side, whilst on the American there are many, and among them, preëminently, the Orinoco and the Amazon, these vast rivers carrying their detritus far out to sea, and helping to produce the configuration of the ocean-bottom in question. However this may be, it is doubtless through these deep troughs that much of the cold water of the North Polar current finds its way.

In accordance with this, we perceive, on examining the temperature of the water, after the African verge of the greater or eastern sea trough is reached, that there is a difference in temperature between the surface and that at a depth of

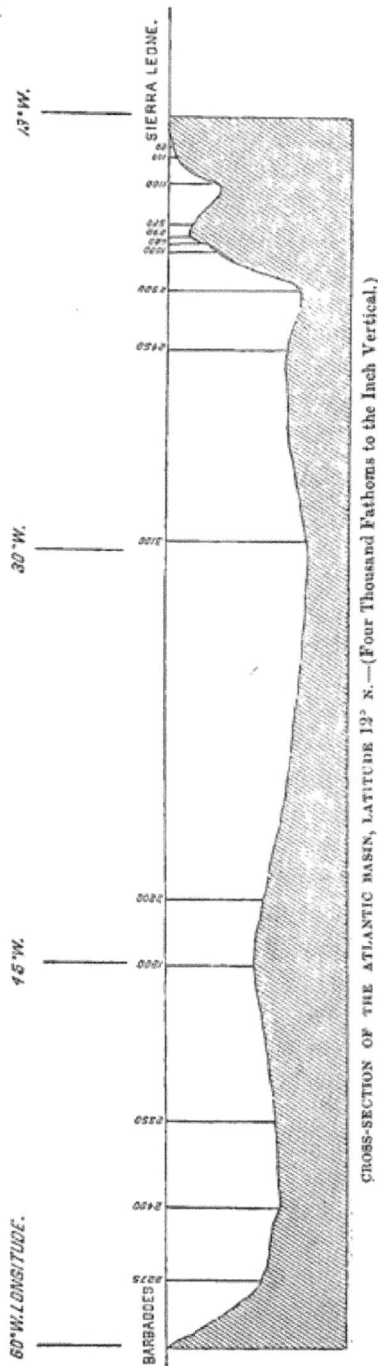

CROSS-SECTION OF THE ATLANTIC BASIN, LATITUDE 12° N.—(Four Thousand Fathoms to the Inch Vertical.)

SIERRA LEONE.

BARBADOES.

HAVANA.

85.°W. LONGITUDE. 75.° 65.° 55.° 45.° 35.° 25.° 15.°

90°F.
80°
70°
60°
50°
30°
20°

450 FATHOMS.

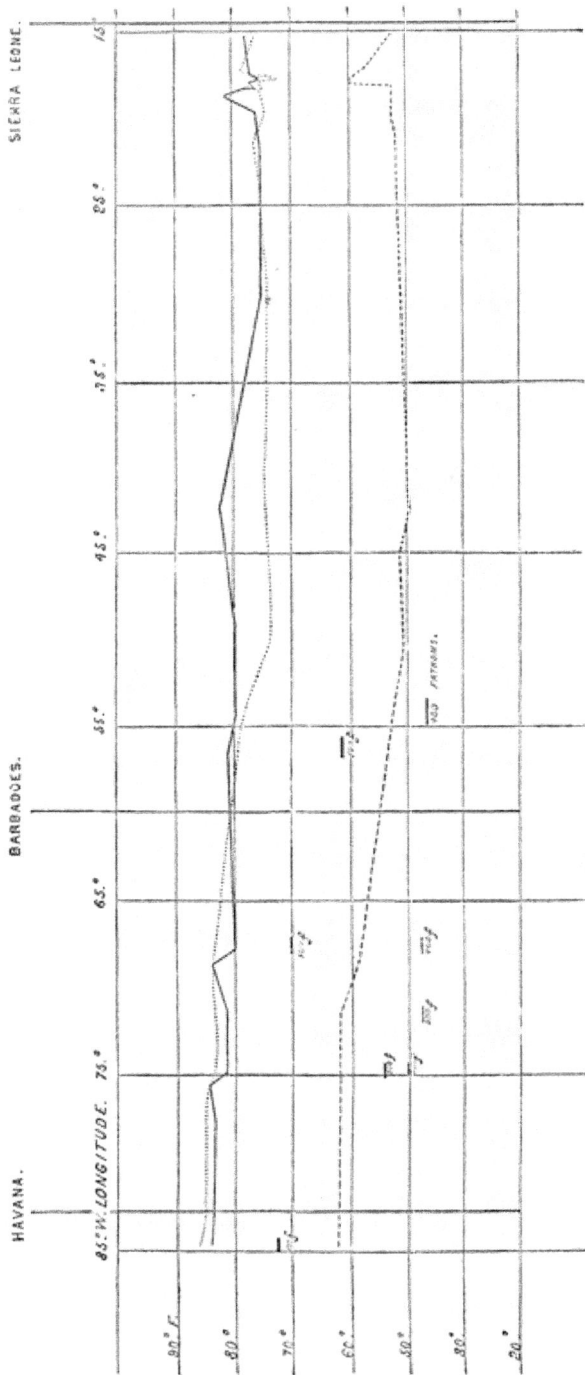

DIAGRAM OF THE TEMPERATURE OF THE AIR, SURFACE OF THE WATER, AND DEEP WATER.

not more than two hundred fathoms, exceeding twenty-five degrees in many cases. This decline of temperature increases as the depth increases, one observation giving an additional fall of four degrees at an additional depth of 200 fathoms. It is not, however, intended to affirm that the mass of cold water is restricted to these deep troughs, since even in the West India seas, at similar depths, low temperatures are observed, and this, though the heat of the surface-water has become very much higher. In those seas, while the surface temperature was 84°, the thermometer, at depths of 400 and 500 fathoms, marked 48°, and these, it must be remembered, were the indications of an uncompensated instrument, which was bearing a pressure of at least half a ton on each square inch of its surface, and hence registering degrees that were higher than the truth. This accords with the observation of Mr. Barrett, that in the deepest part of the sea, near Jamaica, there exists a temperature not far above that of the freezing point of fresh water.

The foregoing diagram represents the temperatures of the air, those of the surface of the water and at different depths, usually, however, at 200 fathoms, from Sierra Leone to Barbadoes, and from Barbadoes to Havana. The solid line represents the temperatures of the air, the dotted line that of the surface of the water, the broken line that of the water at 200 fathoms. Here and there, as will be easily recognized, isolated observations at 100, 300, 400, and 500 fathoms have been introduced. The vertical lines indicate longitudes, the degrees of temperature are seen at the side.

This diagram shows that the air line, after its oscillations near the African coast are over, exhibits a general rise to Barbadoes and thence to Havana. There is, however, a marked departure from the gradual character of this rise at a distance of about two-thirds of the way across. The surface temperatures are for the most part lower than the air temperatures, until the West India seas are reached, when they become higher. They do not exhibit that sudden rise observed in the air temperatures at 45° west longitude. The line of deep-water temperatures is, of course, lower than that of the surface all the way across, and exhibits the same contour in a general manner as the surface-line, rising as it comes into the Caribbean Sea.

From Table IV. it appears that the specific gravity of the water becomes greater as the location is further west. This conclusion I found to be confirmed by an examination of the solid residue left after evaporation to dryness of different samples. I need not quote many of these confirmatory experiments. It will be understood that the customary care was taken in ascertaining the weight of these solid residues to avoid hygrometric complication and the escape of their more volatile ingredients. As an example, it may be mentioned that surface water taken near the African coast on March 1, gave 39.600 grammes of solid residue in 1,000 grammes of water, and another sample of surface-water taken on April 10, in the Caribbean Sea, gave 40.520 grammes in 1,000 grammes. It is probable that the concentration here remarked was due to the drying effect of the north-east trade wind.

The specific gravity of the surface-water is of course affected by the rate at which evaporation is taking place, a concentration ensuing as evaporation goes on, and the density becoming greater. There is, therefore, a tendency in such concentrated waters to leave the surface and pass downward, until they reach a region in which the density corresponds to their own. These considerations have led some persons to suppose that this descent would be continued to very great depths, and that this movement must sensibly affect the general motion taking place in the waters of the sea. If, however, we compare the density of the surface-waters in the tropical Atlantic as I have represented them in the foregoing tables, with the density of surface-water in higher latitudes, where both the air and water are much colder, and where evaporation is much less active, as shown, for instance, in latitudes between 50° and 60° north in the cruises of Dr. Carpenter, we see how small the difference is. When we remember, also, how slight a variation of temperature, as has been already said, completely masks these differences arising from concentration, we may safely conclude that the cause of disturbance may be overlooked.

25

In the adjoining diagram the vertical lines represent longitudes, the horizontal lines specific gravities, the solid line specific gravities of the surface-water, the dotted specific gravity of water at two hundred fathoms. On comparing the two latter lines together, we see that they almost coincide; and again, on comparing this diagram with that of the temperatures and also with the cross section of the Atlantic basin, a curious resemblance will be readily detected, viz., that they all exhibit an elevation about the meridian of 45° west longitude, as though the contour of the bottom was not without influence on the physical state of the water above it.

The general conclusion which may be drawn from these results as to temperatures and specific gravities is, that there exists all over the bottom of the tropical Atlantic and Caribbean Sea a stratum of cold water, —cold, since its temperature is below 50°. This is the conclusion to which Dr. Carpenter has come, as respects the Atlantic in higher north latitudes; and in this important particular the cruise of the Mercury must be considered as offering confirmatory proofs of the correctness of the deductions drawn from the cruises of the Lightning and Porcupine.

There are reasons for supposing that, so far from this water being

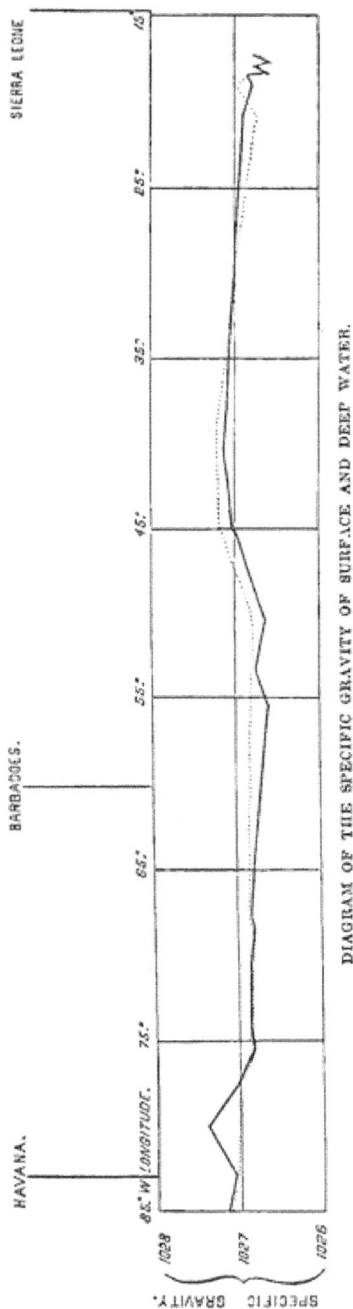

DIAGRAM OF THE SPECIFIC GRAVITY OF SURFACE AND DEEP WATER.

stagnant, its whole mass has a motion toward the Equator, whilst the surface-waters in their turn have a general movement in the opposite direction.

As the samples of water had already been kept for some months before they came into my possession, though they had been confined, as has been said, in bottles closed by corks covered with sealing-wax, their gaseous ingredients must have undergone much change. I have already pointed out the disturbance that necessarily ensues in these gaseous ingredients as they are being drawn to the surface, and another disturbance occurs when they are brought more fully in contact with the air on being poured into the bottles. Interchange by diffusion then rapidly takes place, tending to make their proportions approach those of surface-water. For these reasons I did not consider it necessary to make an analysis of the gases afforded by these specimens. Dr. Carpenter, from his experiments, concluded that when freshly-drawn water is tested, the proportion of carbonic acid increases as the stratum of water is from deeper sources.

I made some examinations of the organic matter contained in these waters, both by incinerating the solid residue and by the permanganate test. It has been customary to divide such organic ingredients into two groups, the decomposed and decomposable, in investigations respecting the sustenance of animal life at great ocean depths. To estimate the relative proportion of these groups, a sample of the water is divided into two parts—one is acidified, and then to both an excess of a standard solution of permanganate of potassa is added. After three hours, iodide of potassium and starch are introduced to check further reaction, and the excess of permanganate in each sample is then estimated by a standard solution of hyposulphite of soda. From the portion to which free acid was added the decomposed and easily-decomposable organic matter can be estimated, and from the other the decomposed alone. I resorted to this method of examination in several instances, but was discouraged from prosecuting it, for such reasons as have already been stated in the foregoing case of the gaseous constituents. It needed no special proof that organic matter was present in every one of these samples, for the clearest of them contained shreddy and flocculent material, some of them quantities of sea-weed in various stages of decomposition. With these vegetable substances were the remains of minute marine animals. As bearing upon this subject, I found, on incinerating the solid residue of a sample of water taken from two hundred fathoms, that the organic and volatile material was not less than eleven per cent. of the whole. Though the quantity of organic substance diminished as the stratum under examination was deeper, there still remained a visible amount in the water of four hundred or five hundred fathoms. It is probable, therefore, that even at the bottom of the ocean such organic substance may exist, not only in solution affording nutriment to animals inhabiting those dark abysses, as Professor Wyville Thompson has suggested, but also in the solid state. Plants of course cannot grow there, on account of the absence of light.

In order to determine whether any hitherto-unknown element existed in these waters, I subjected the solid residue to examination with the spectroscope, volatilizing the substances by the aid of a voltaic current and induction coil. A careful examination did not reveal the presence of any spectral lines, other than those belonging to the well-known elementary substances in sea-water.

The specimens of the bottom, obtained by attaching to the sounding-line quills or wooden tubes, I have transmitted to Dr. Carpenter, who has kindly consented to examine them.

In a letter dated August 10, 1871, recently received, he says : "As far as I can see, they consist of the ordinary Atlantic mud, chalk in process of formation, with the ordinary types of deep-sea foraminifera." The chemical composition of this mud, according to Mr. Forbes, as taken from soundings at 1,443 fathoms in the North Atlantic, is:

Carbonate of lime	50.12
Alumina (soluble in acids)	1.33
Sesquioxide of iron (soluble in acids).	2.17
Silica (in a soluble condition).	5.04
Fine insoluble gritty sand (rock debris)	26.77
Water	2.90
Organic matter	4.19
Chloride of sodium and other soluble salts	7.48
	100.00

All of which is respectfully submitted.

HENRY DRAPER, M.D.

SCHOOL-SHIP MERCURY, }
NEW YORK, April 29, 1871. }

SIR—I would respectfully submit to your Honorable Board the following report of all reliable deep-sea soundings made from this ship during her late practice cruise, together with a description of the manner in which they were made. The soundings were taken within the parallels of eleven and thirteen north latitude, from the coast of Africa, in the vicinity of Sierra Leone, to the Island of Barbadoes, W. I., through the Caribbean Sea to the west end of Cuba. The instrument used was a detaching apparatus, invented by the late Lieutenant J. M. Brooke, U. S. N., and supplied to this ship through the kindness of Commodore James Alden, Chief of the Bureau of Navigation, Navy Department, Washington. I found it worked admirably, never failing to bring up specimens of the bottom, except when coming in contact with rocky bottom. The quills inserted in the cell or holder, as seen in the accompanying sketch of the apparatus, preserve the specimens in better condition for microscopic examination than the ordinary "arming" of soap or tallow would bring up.

Two spherical shot, weighing thirty-two pounds each, were used at every cast as a sinker, the sounding-line made of cotton, one-seventh of an inch in diameter, capable of sustaining one hundred and sixty pounds in the air; the cord carefully waxed to overcome, as much as possible, the resistance of the line from friction of the water. All soundings were made from a boat, as repeated trials satisfied me that no reliance could be placed on those made from the ship, on account of the difficulty of keeping the vessel in position and the line "up and down." A boat, fitted expressly for the purpose, with a reel holding four thousand fathoms, working on friction rollers, the line passing over the bow through a leader, thus enabling the boat, with the aid of oars, to be kept in position and the line perpendicular on the shot, thus determining to a certainty when the sinker reached the bottom.

Temperature and specimens of sea-water at various depths were obtained by a self-registering metallic thermometer carefully adjusted; all specimens secured have been carefully preserved, and now await your order. All soundings were made under my immediate supervision, and the position of the ship at the time of sounding carefully noted, for which I respectfully refer you to the accompanying table.

Very respectfully, your obedient servant,

P. GIRAUD, *Captain.*

ISAAC BELL, Esq., *President, etc.*

On the 11th of April attached self-registering thermometer twelve feet above shot. Coming in contact with bottom, broke the "mercury" one.

Very respectfully, your obedient servant,

P. GIRAUD, *Captain.*

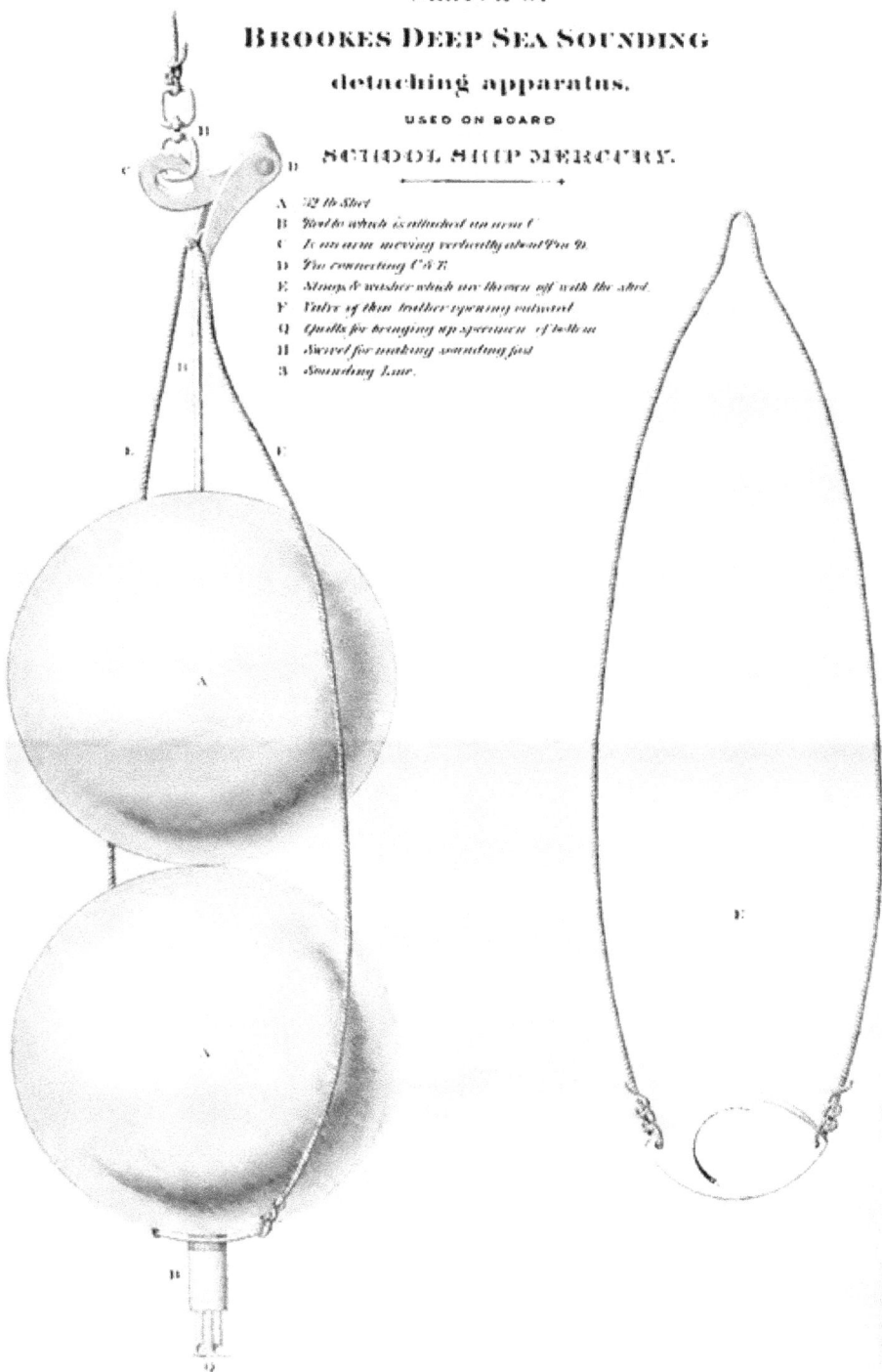

SKETCH OF
BROOKES DEEP SEA SOUNDING
detaching apparatus.
USED ON BOARD
SCHOOL SHIP MERCURY.

A   32 lb Shot
B   Rod to which is attached an arm C
C   Is an arm moving vertically about Pin D
D   Pin connecting C & F
E   Strap & washer which are thrown off with the shot
F   Valve of thin leather opening outward
G   Quills for bringing up specimen of bottom
H   Swivel for making sounding fast
I   Sounding Line.

STATEMENT *of all reliable Deep Sea Soundings, made from New York Nautical School-ship "Mercury" during Practice Cruise,* **1871**, *from Sierra Leone to Barbadoes and to New York. Used Brooks' Detaching Apparatus, with two thirty-two pound shot. Sounding line used, cotton cord, one-seventh of an inch in diameter. All soundings made from boat.*

DATE.	Position of Ship at Time of Sounding. Latitude.	Position of Ship at Time of Sounding. Longitude.	Fathoms.	Temperature of Air.	Temperature of Surface Water.	Various Depths.	Temperature at Various Depths.	Obtained Specimens of Bottom (*).	Current.		Time of Sounding.	REMARKS.
						Fathoms						
February 23	8° 50′	15° 47′	1,100	79°	78°	200	54°	......	SW	½ knot.	2 P.M.	
" 25	9° 15′	17° 12′	520	78°	79°	200	58°	*	SSW	½ "	3 "	
" 26	10° 05′	17° 25′	200	76°	74°	200	60°	*	S by W	½ "	3 "	
" 27	10° 42′	17° 46′	680	78°	77°	200	60°	*	S	½ "	3 "	
" 28	11° 24′	18° 09′	1,000	77°	76°	200	53°	*	S by W	½ "	3 "	
March 1	11° 26′	18° 20′	......	80°	77°	200	53°	......	......	......	......	
" 2	11° 33′	18° 38′	......	82°	77°	200	53°	......	......	......	......	
" 3	11° 35′	19° 33′	2,900	77°	76°	200	53°	*	SSW	½ knot.	3 P.M.	
" 4	11° 06′	21° 55′	2,450	76°	77°	200	32°	*	S	½ "	3 "	
" 6	11° 32′	29° 13′	3,100	76°	75°	......	......	......	......	......	3 "	Reeling in, parted line; lost 2,200 fathoms.
" 10	12° 38′	42° 31′	2,200	85°	76°	200	50°	*	SW	½ knot.	2 "	
" 11	13° 03′	44° 51′	1,900	84°	76°	200	51°	*	SW by W	¼ "	2 "	
" 13	13° 06′	50° 38′	2,250	80°	75°	200	51°	*	WSW	½ "	11 A.M.	
" 14	13° 08′	53° 48′	2,400	80°	78°	400	47°	*	WSW	½ "	2 P.M.	
" 15	12° 55′	W. 56° 46′	2,275	81°	80°	100	62°	*	W	½ "	2 "	
April 4	17° 13′	67° 29′	2,500	80°	84°	100 400	70° 48°	*	W	½ "	3 "	
" 5	17° 08′	68° 48′	2,680	84°	84°	200	50°	*	W	½ "	3 "	
" 6	17° 09′	71° 47′	......	82°	84°	500 200	48° 62°	......	W by N	½ "	4 "	
" 7	17° 27′	74° 33′	1,500	82°	84°	400 500	50° 54°	*	W by N	½ "	4 "	
" 8	18° 11′	76° 00′	950	85°	85°	200	62°	Rocky	WNW	½ "	3 "	
" 9	19° 12′	77° 54′	......	84°	85°	200	62°	......	W by N	½ "	3 "	Wind, ENE.
" 11	20° 48′	83° 02′	1,900	84°	85°	200	62°	......	SW	19 knots	4 "	Wind, E to S.
" 13	23° 19′	84° 17′	1,240	84°	86°	200 100	62° 72°	......	NW	1 knot.	5 "	Wind, NE by E.

On the 11th of April attached self-registering thermometer twelve feet above shot. Coming in contact with bottom, broke the "mercury bulb."

Very respectfully, your obedient servant,

P. GIRAUD, *Captain.*

ABSTRACT *Log of Nautical School-ship "Mercury."—Department of Public Charities and Correction—1870-1871.*

DATE	TEMPERATURE OF AIR AT			TEMPERATURE OF WATER AT			BAROMETER AT			Latitude at Noon.	Longitude at Noon.	Courses.	Distances.	Winds.	REMARKS.
	Midnight.	Noon.	8 P.M.	Midnight.	Noon.	8 P.M.	Midnight.	Noon.	8 P.M.						
Dec. 22	40°	42°	41°	50°	54°	56°	30.10	30.14	30.12	40° 04′	69° 59′	S 62° E	195′	WNW to SW	From Hart's Island.
" 23	43°	48°	41°	51°	52°	50°	30.11	30.18	30.18	38° 38′	66° 03′	S 74° E	197′	NW to NNE	Rough sea, ship rolling heavily.
" 24	54°	65°	63°	63°	61°	65°	30.05	30.10	30.11	37° 49′	64° 00′	N 87° E	192′	NW, W, S	Rough sea, ship rolling heavily.
" 25	54°	54°	54°	64°	66°	68°	29.35	29.25	29.25	38° 05′	61° 15′	S 87° E	140′	NW, W, WNW	Fresh gale, high sea.
" 26	58°	59°	57°	67°	65°	67°	29.40	29.45	29.50	38° 04′	56° 57′	N 87° E	211′	SW, W, NNW	Heavy gale, accompanied by furious squalls.
" 27	58°	59°	57°	65°	67°	68°	29.40	29.42	29.60	55° 40′	51° 55′	S 89° E	217′	NNW, N, NNE	Light breeze, confused sea, sky overcast.
" 28	58°	61°	60°	67°	65°	68°	29.70	29.80	29.85	37° 20′	49° 56′	S 79° E	183′	SW, SSW, WSW	Moderate breeze and pleasant weather throughout.
" 29	59°	68°	68°	66°	66°	68°	29.92	29.98	30.05	37° 07′	46° 05′	S 89° E	190′	SW, NE	Current ENE twelve miles, sky overcast.
" 30	62°	68°	66°	62°	64°	66°	30.32	30.32	30.42	36° 20′	43° 08′	S 48° E	165′	SSE to ESE	Moderate breeze and pleasant.
" 31	64°	62°	62°	63°	65°	63°	30.28	30.05	30.12	36° 40′	42° 15′	N 54° E	166′	SSE, S, SSW	Moderate breeze and pleasant, passing clouds.
Jan. 1	62°	63°	60°	65°	62°	64°	30.40	30.36	30.57	35° 42′	40° 02′	S 84° E	189′	SW to SSW	Moderate breeze and fine weather.
" 2	61°	62°	67°	64°	64°	64°	30.18	0.12	30.11	37° 29′	36° 15′	S 77° E	90′	SW, WSW, SW	Fine weather, light passing clouds.
" 3	70°	68°	68°	64°	63°	63°	30.18	29.42	30.40	37° 29′	34° 04′	N 78° E	61′	SSW, S, SW, SE	Light airs, accompanied by squalls.
" 4	66°	68°	62°	63°	64°	65°	30.40	29.42	30.46	34° 50′	30° 47′	S 46° E	55′	SW to S	Clear and pleasant.
" 5	65°	68°	64°	65°	66°	64°	30.40	29.38	30.42	24° 59′	29° 55′	S 50° E	120′	SE, light	Fine weather.
" 6	64°	64°	64°	66°	65°	65°	30.42	29.42	30.46	34° 21′	31° 03′	N 29° W	45′	SE, N by S	Moderate breeze and cloudy.
" 7	61°	61°	60°	64°	65°	60°	30.48	29.48	30.53	34° 12′	31° 01′	N 14° W	140′	Easterly	Fresh breeze and squally, sea rough.
" 8	60°	68°	63°	62°	66°	62°	30.52	29.52	30.51	35° 13′	30° 18′	N 11° E	147′	E to SW	Moderate breeze, heavy passing clouds.
" 9	68°	61°	60°	63°	65°	60°	29.54	29.50	30.50	35° 47′	29° 70′	N 19° E	99′	E, SE to S	Moderate breeze and cloudy, sea rough.
" 10	63°	68°	60°	63°	63°	61°	30.46	30.44	30.42	39° 18′	28° 50′	N 16° E	72′	ESE, light	Moderate breeze and fine weather.
" 11	57°	59°	59°	60°	62°	60°	30.22	30.28	30.24	40° 26′	27° 38′	N 19° E	64′	E to S, light	Light and baffling, cloudy.
" 12	62°	61°	62°	61°	61°	60°	30.14	30.09	30.09	40° 47′	27° 29′	N 67° E	58′	SW, light	Fine weather.
" 13	58°	58°	58°	59°	59°	58°	29.98	29.93	30.12	40° 28′	26° 22′	N 86° E	92′	S	Sky overcast, sea smooth.
" 14	64°	65°	63°	61°	65°	62°	29.98	29.94	29.93	55° 54′	22° 22′	S 45° E	230′	SW to NW	Fresh breeze and squally.
" 15	65°	60°	63°	62°	64°	65°	29.94	29.94	29.98	33° 50′	19° 50′	S 39° E	200′	NW	Fresh breeze, passing clouds, rough sea, 88b current one knot per hour.
" 16	60°	59°	64°	63°	64°	64°	29.95	29.94	29.94	33° 31′	17° 20′	S 45° E	117′	NW	Heavy passing squalls, squally, rolling swell NW.
" 17	64°	68°	63°	61°	65°	63°									First part fine weather, second part squally and rainy. At 9 A.M. anchored at Funchal, Madeira, lat. 32° 38′, long. 16° 55′.

Date						Barometer			Lat.	Long.	Wind	Remarks
18	78°	78°	68°	70°	68°		29.90	29.92	29.93				Lying at anchor at Funchal, Ma. Daily average of barometer, 30.10. Enjoyed fine weather.
19	68°	68°	68°	70°	67°		29.90	29.88	29.88			NE, light	Sailed for Las Palmas, by Canary, at 8 P.M. Smooth sea and pleasant weather.
20	65°	71°	66°	70°	67°		29.63	29.68	29.61			NE, light	Sea smooth.
21	65°	65°	65°	70°	64°					16° 49'	S 81° E	NE	At 10 A.M. anchored at Las Palmas, lat. 28° 07', long. 15° 25'.
22	59°	59°	63°	64°	59°		29.57	29.56	29.55	15° 53'	S 13° E		Lying at anchor at Las Palmas. During our stay we experienced unusually cool weather, rough sea and stormy weather from NE, to which the harbor is exposed.
23	59°	58°	63°	63°	59°		30.06	30.08	30.08				
24	63°	63°	63°	63°	59°		30.04	30.02	30.03				
25	64°	64°	64°	62°	62°		29.60	29.66	29.57				
26	59°	60°	63°	63°	62°		29.99	29.98	29.96				
27	57°	59°	60°	61°	62°		30.01	30.04	30.08				
28	62°	62°	63°	63°	63°		30.02	30.01	30.00				
29	63°	63°	63°	64°	64°		29.98	29.96	29.98				
30	65°	66°	64°	64°	62°		29.99	29.95	29.89				
31	64°	64°	65°	65°	62°		29.76	29.85	29.78				
Feb. 1	66°	66°	66°	68°	60°		29.88	29.84	29.85	27° 48'	S 66° E	Baffling S	Sailed for Sierra Leone at 3 P.M. Pleasant weather.
2	63°	64°	66°	65°	65°		24.24	24.24	24.28	28° 30'	S 9° W	SW to N	Light breeze, passing clouds, pleasant weather.
3	67°	67°	66°	70°	70°		29.99	29.96	30.00	29° 23'			
4	68°	69°	67°	69°	66°								
5	65°	65°	66°	68°	64°								
6	68°	68°	64°	72°	68°		29.99	29.94	29.99	21° 19'	S 50° W	NNE to N	Fresh breeze, rough sea, sky clear. Current, SW⅜S, 15 miles.
7	63°	66°	63°	70°	66°		29.84	29.80	29.73	20° 06'	S 33° W	NE	Fresh breeze, pleasant weather. Current, SW⅛W, 21'.
8	66°	69°	66°	72°	72°		29.76	29.73	29.70	19° 42'	S 8° E	NE	Fresh breeze, pleasant weather, light passing clouds. Current, SSW, 17 miles.
9	72°	72°	73°	73°	72°		29.70	29.65	29.69	18° 52'	S 12° E	E by N	Fresh breeze, rough sea. Current, S⅛W, 14 miles.
10	74°	75°	73°	79°	75°		29.69	29.63	29.64	17° 41'	S 35° E	NE	Fresh breeze, smooth sea. Current, S⅜E, 13 miles.
11	77°	78°	77°	83°	78°		29.63	29.60	29.63	15° 36'	S 62° E	NE	Pleasant weather, smooth sea. Current, SE, 10 miles.
12	79°	80°	78°	82°	79°		29.62	29.70	29.62	15° 00'	S 80° E	Light air & calm	Atmosphere very hazy. Current, E, 8 miles.
13	79°	83°	80°	83°	79°		29.64	29.64	29.70	14° 26'	S 69° E	Light air & calm	Atmosphere very hazy. Current, SE, 15 miles.
14	80°	85°	79°	84°	80°		29.60	29.60	29.60				At 4 P.M. anchored at Sierra Leone. Latitude, 8° 30' N; longitude, 13° 18' W.
15	80°	83°	79°	85°	81°		29.62	29.60	29.63				Lying at anchor at Sierra Leone. Land and sea breeze, occasionally interrupted by calms, hazy atmosphere, and pleasant weather.
16	81°	81°	84°	83°	82°		29.58	29.38	29.62				
17	82°	83°	79°	85°	83°		29.63	29.59	29.70				
18	82°	82°	73°	83°	81°		21.70	21.61	21.63				
19	82°	82°	81°	82°	81°		29.64	29.66	29.68				
20	78°	79°	80°	81°	81°		29.66	29.66	29.64				
21	79°	79°	80°	83°	82°		29.70	29.68	29.70	9° 11'	N 62° W	SW to NE	Sailed for Barbadoes at 8 A.M. Weather hazy, wind light. Current, S, 8 miles.
22										14° 39'			
23	80°	82°	79°	81°	83°		29.76	29.70	29.74	15° 47'	S 75° W	W to SE, light	Weather hazy, wind light. Current, SSW, 7 miles.

DATE	TEMPERATURE OF AIR AT			TEMPERATURE OF WATER AT			BAROMETER AT			Latitude at Noon	Longitude at Noon	Course	Distances	Winds	REMARKS
	Midnight	Noon	8 P.M.	Midnight	Noon	8 P.M.	Midnight	Noon	8 P.M.						
Feb. 24	78°	81°	79°	78°	80°	78°	29.70	29.68	29.66	4° 09'	16° 21'	N 66° W	42'	W to N	Weather hazy, wind light. Current, southerly.
,, 25	78°	84°	80°	78°	80°	78°	29.68	29.66	29.70	6° 13'	17° 06'	N 83° W	47'	NW to NE	Very hazy, wind light. Current, SSE, 8 miles.
,, 26	77°	82°	76°	77°	79°	76°	29.74	29.74	29.74	9° 53'	17° 31'	N 25° W	46'	Light air & calm	Very hazy, wind light. Current, SSE, 8 miles.
,, 27	74°	82°	77°	74°	78°	75°	29.76	29.76	29.76	10° 36'	17° 48'	N 19° W	43'	Light air & calm	Very hazy, wind light. Current, SSE, 8 miles.
,, 28	74°	78°	78°	75°	76°	76°	29.74	29.68	29.64	11° 28'	18° 06'	N 18° W	58'	NW, light	Very hazy, wind light.
Mar. 1	73°	80°	73°	76°	73°	76°	29.68	29.66	29.66	11° 36'	18° 34'	S 84° W	17'	W, light & baff'g.	Atmosphere hazy, peculiar to the African coast.
,, 2	74°	80°	78°	73°	77°	76°	29.68	29.68	29.68	11° 39'	18° 33'	N 37° W	20'	W, light & baff'g.	Atmosphere hazy, peculiar to the African coast.
,, 3	74°	80°	70°	74°	76°	70°	29.70	29.70	29.70	11° 33'	19° 27'	S 84° W	52'	WNW to N	Light and calm, sky clear. Current, S by E, 12 miles.
,, 4	73°	77°	71°	73°	78°	71°	29.72	29.74	29.72	11° 08'	21° 46'	S 80° W	141'	Northerly	Moderate breeze, clear. Current, SBW, 15 miles.
,, 5	74°	80°	76°	74°	78°	76°	29.72	29.73	29.74	11° 05'	26° 00'	S 87° W	260'	NNE	First of the NE trades, fresh. Current, WSW, 6 miles.
,, 6	74°	79°	74°	74°	73°	74°	29.80	29.78	29.78	11° 31'	29° 02'	N 83° W	222'	NE	Fresh breeze and passing clouds.
,, 7	74°	79°	75°	74°	76°	75°	29.78	29.76	29.76	11° 46'	32° 23'	N 86° W	202'	NE by E	Fresh trades, fine weather, sea rough.
,, 8	72°	78°	77°	74°	75°	76°	29.78	29.76	29.76	12° 01'	35° 48'	N 83° W	203'	NE by E	Fresh trades, fine weather, sea rough.
,, 9	74°	80°	76°	74°	76°	76°	29.76	29.72	29.72	12° 15'	38° 64'	N 87° W	194'	NE	Moderate breeze, sky overcast.
,, 10	75°	80°	76°	75°	76°	76°	29.76	29.76	29.76	12° 32'	42° 24'	N 84° W	197'	NE by E	Moderate breeze, passing clouds.
,, 11	76°	81°	81°	76°	77°	77°	29.74	29.73	29.72	13° 01'	44° 47'	N 89° W	185'	NE by E	Moderate breeze, passing clouds.
,, 12	76°	81°	80°	76°	77°	76°	29.74	29.74	29.74	13° 07'	47° 45'	N	181'	ENE	Steady trades, passing clouds.
,, 13	78°	81°	78°	76°	78°	78°	29.78	29.76	29.76	13° 08'	50° 38'	W	109'	ENE	Steady trades, passing clouds.
,, 14	78°	81°	81°	77°	80°	79°	29.82	29.81	29.82	13° 08'	53° 44'	N 88° W	182'	ENE	Steady trades, passing clouds. Current.
,, 15	74°	79°	77°	74°	75°	74°	29.80	29.80	29.74	12° 54'	56° 36'	S 84° W	174'	ENE	Steady trades, passing clouds. Current, ½ a knot, W. 14 miles, W.
,, 16	79°	83°	82°	82°	83°	82°	29.78	29.78	29.80						Made Barbadoes at 8 o'clock A.M., thirty miles distant. At 12 o'clock M. came to anchor at Bridgetown. Lat. 13° 04', long. 59° 38.
,, 17	77°	81°	80°	78°	81°	79°	29.80	29.80	29.81						Lying at anchor at Bridgetown. Steady NE trades and pleasant weather.
,, 18	76°	80°	80°	78°	80°	80°	29.81								
,, 19	71°	78°	75°	76°	78°	77°	Daily average while at Bridgetown, Barbadoes, 29.80.								
,, 20	74°	78°	73°	74°	77°	73°									
,, 21	76°	80°	79°	76°	79°	79°									
,, 22	73°	84°	83°	73°	83°	83°									

33

Date							Bar.				Temp.	Lat.	Long.	Wind		Dist.	Wind	Remarks
"							29.81	29.80	29.80			14° 58'	62° 10'	N 64° W	172'		F. to NE	Sailed for St. Thomas at 10 o'clock A.M. Moderate breeze, fine weather. Current, WNW, 12'.
"	26						29.90	29.80	29.80			17° 01'	63° 31'	N 35° W	167'		NE	Moderate breeze, fine weather. Current, W, 12'.
"	27						29.80	29.82	29.82									Anchored at St. Thomas at 1 o'clock P.M. Lat. 18° 21', long. 64° 58'.
"							29.64	29.82	29.82									Lying at anchor at St. Thomas. Light baffling winds, heavy passing clouds, and pleasant weather.
April 1																		Daily average while at St. Thomas, 29.82.
"							29.78	29.80	29.80			17° 14'	67° 96'	S 62° W	153'		ESE to SSE	Sailed for New York at 8 o'clock A.M. Sky overcast, breeze light. Current, W.
"							29.80	29.80	29.80			17° 08'	68° 40'	S 85° W	79'		ESE	Breeze light. Current, W.
"							29.80	29.76	29.76			17° 05'	71° 55'	S 85° W	157'		NE	
"							29.76	29.78	29.78			17° 25'	74° 05'	N 82° W	154'		NE to E	
"							29.80	29.86	29.86			18° 09'	75° 33'	N 69° W	117'		ENE	
"							29.80	29.82	29.82			19° 00'	77° 41'	N 80° W	113'		NE to E	
"							29.76	29.76	29.72			20° 11'	80° 08'	N 62° W	145'		NE	Current, WSW, 11'
"							29.74	29.76	29.78			22° 44'	82° 46'	N 73° W	156'		E to S	Current, SW, 19'.
"							29.80	29.80	29.76			21° 42'	84° 57'	N 59° W	108'		E to SE	Current, WSW, ¾ of a knot.
"	13						29.80	29.80	29.98			23° 14'	84° 14'		127'		ENE, light	Passed Cape St. Antonio, west end of the Island of Cuba, at 6 o'clock P.M., distant 4'. Current, NNW, ½ a knot. Current, NE by ¼, 3 miles.
"	14						29.64	29.62	29.52			25° 35'	81° 27'	N 80° E	145'			Fresh gales from SW, high sea. Current, N by E, 3½ knots.
"	15						29.55	29.42	29.42			26° 12'	79° 22'	N 37° E	243'		SSW, moderate.	
"	16						30.32	2.70	2.70			28° 04'	79° 08'	N 8° E	172'		SW	
"	17						29.90	29.90	29.90			30° 18'	80° 06'	N 36° W	93'		NE to NNW	Long rolling swell from the north. Current, NNE, 2½ knots.
"	18						29.80	29.82	29.80			32° 09'	79° 05'	N 25° E	122'		W to SW	Strong tide-rips, atmosphere hazy. Current, NE by N, 2 knots.
"	19						29.90	29.92	29.80			34° 31'	74° 52'	N 57° E	250'		SW	Fresh breeze with rough sea, sky overcast. From 7.45 P.M. to 7.35 P.M. passed out of the Gulf-stream, during which time the temperature of water changed from 81° to 62°.
"	20						29.66	29.58	29.58			38° 30'	74° 10'	N 10° E	240'		SSW	Batting winds accompanied with squalls, passing clouds, and hazy.
"	21						29.53	29.56	29.70			40° 27'	73° 52'	N 5° E	130'		S to W	At 8 A.M. signalled New York pilot from pilot-boat "Mary A. Williams" No. 13. Arrived off Sandy Hook at noon.

Very respectfully submitted,

P. GIRAUD, *Captain.*

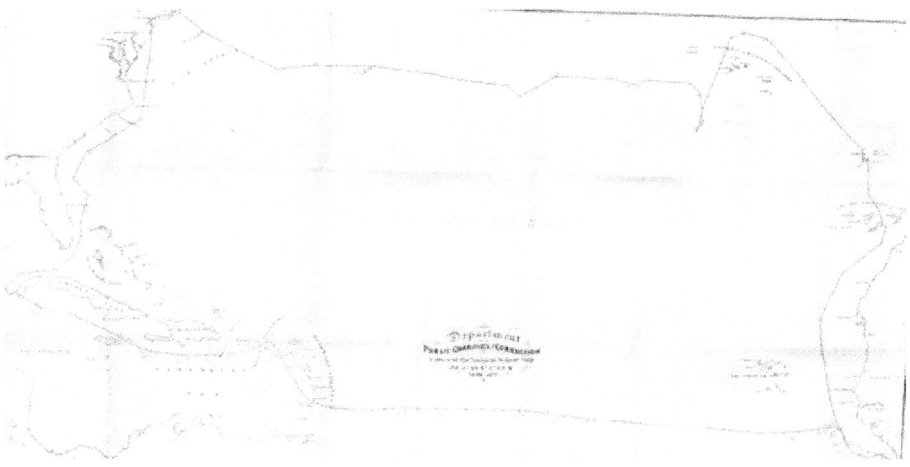

Department of Public Charities and Correction.

CRUISE

OF

CHOOL-SHIP "MERCURY"

IN

TROPICAL ATLANTIC OCEAN.

1870—1871.

NEW YORK:

THE NEW YORK PRINTING COMPANY,
Nos. 81, 83, and 85 Centre Street.

1871.

www.ingramcontent.com/pod-product-compliance
Lightning Source LLC
Chambersburg PA
CBHW021441090426
42739CB00009B/1591